DATE DUE

OCT 2 3 2007			
GAYLORD			PRINTED IN U.S.A.

STRUCTURES

Please visit our web site at: www.garethstevens.com
For a free color catalog describing Gareth Stevens Publishing's list of high-quality books and multimedia programs, call 1-800-542-2595 (USA) or 1-800-387-3178 (Canada). Gareth Stevens Publishing's fax: (414) 332-3567.

Library of Congress Cataloging-in-Publication Data

Structures.—North American ed.
    p. cm. — (Discovery Channel school science: physical science)
  "First published in 2001 as Sky-high: the structures files by Discovery Enterprises, LLC, Bethesda, Maryland." — T.p. verso.
  Summary: An overview of how challenges to design and construction have been overcome throughout history to make it possible to build structures that can stand up to natural forces.
  ISBN 0-8368-3364-3 (lib. bdg.)
  1. Structural engineering—Juvenile literature. 2. Buildings—Juvenile literature.
[1. Structural engineering. 2. Buildings.] I. Title. II. Series.
TA634.S559   2003
624.1—dc21   2002030535

This edition first published in 2003 by
Gareth Stevens Publishing
A World Almanac Education Group Company
330 West Olive Street, Suite 100
Milwaukee, WI 53212 USA

Writers: Jackie Ball, Justine Ciovacco, Bill Doyle, Kathleen Feeley, Darcy Lockman, Monique Peterson, Tanya Stone.
Editor: Justine Ciovacco.
Photographs: Cover, King Kong, Hulton Getty/Archive Photo; p. 4, skyline, Corel; p. 5, Washington Monument, Corel; p. 8, Golden Gate, Corel; p. 9, Sydney Harbor Bridge, Corel; p. 10, Petronas Towers, © Hugh Sitton/Stone, Akashi Kaikyo Bridge, © Kaku Kurita/TimePix; p. 11, Chunnel, © Michel Spingler/AP Photo, Gateway Arch, Corel; p. 13, Pentagon, © Peter Gridley/FPG International, Corp.; p. 14, Safeco under construction, © Mariners Club; p. 15, finished Safeco Field, © Reuters NewMedia Inc./Corbis; p. 16, Jimi Hendrix, © Joseph Sia/Archive Photos; p. 17, EMP/Space Needle and EMP entrance, © Lara Swimmer/Experience Music Project, EMP aerial, © Stanley Smith/Experience Music Project; pp. 18–19, aerial of CBBT Painet; p. 20,

This U.S. edition © 2003 by Gareth Stevens, Inc. First published in 2001 as *Sky-high: The Structures Files* by Discovery Enterprises, LLC, Bethesda, Maryland. © 2001 by Discovery Communications, Inc.

Further resources for students and educators available at www.discoveryschool.com

Designed by Bill SMITH STUDIO
Creative Director: Ron Leighton
Design: Sonia Gauba, Dmitri Kushnirsky, Jay Jaffe, Eric Hoffsten
Photo Editors: Sean Livingstone, Christie K. Silver
Intern: Chris Pascarella
Art Buyer: Lillie Caporlingua
Gareth Stevens Editor: Betsy Rasmussen
Gareth Stevens Art Director: Tammy Gruenewald

All rights reserved to Gareth Stevens, Inc. No part of this book may be reproduced, stored in a retrieval system, or transmitted in any form or by any means, electronic, mechanical, photocopying, recording, or otherwise, without the prior written permission of the publisher except for the inclusion of brief quotations in an acknowledged review.

Printed in the United States of America

1 2 3 4 5 6 7 8 9 07 06 05 04 03

Brooklyn Bridge, Corel; p. 20, John Roebling, Brown Brothers, Ltd.; p. 21, Washington Roebling and Elizabeth Roebling (both), Brown Brothers, Ltd.; p. 21, construction of bridge, © Collection of The New-York Historical Society; p. 22, astronaut, NASA, space station, NASA; p. 23, illustration of completed ISS, NASA; p. 26, worker, © Lewis W. Hine/Archive Photo; p. 27, Empire State Building under construction, © Lewis W. Hine/Archive Photo, skyline with Empire State Building, Corel; p. 30, Petronas Towers, © Kendall LaMontagne/Discovery Communications, Inc.; p. 31, Akashi Kaikyo, © Kaku Kurita/TimePix.

Illustrations: pp. 24–25, icons for map, Jessica Wolk Stanley.
Acknowledgments: p. 26, Paul Starrett excerpt from BUILDING THE EMPIRE STATE. Carol Willis, ed. W. W. Norton & Co., 1998; p. 27, Richmond Shreve excerpt from THE EMPIRE STATE BUILDING: THE MAKING OF A LANDMARK, by John Tauranac. Scribner, 1995.

CONTENTS

Every structure starts with an idea. Turning that idea into a reality makes the design process a challenge for architects and engineers. All structures must stand up to the natural forces of gravity, weather, and occasionally earthquakes or other natural disasters.

The bridges, skyscrapers, monuments, tunnels, and other structures featured in this book have brought pride and problems to the people who designed and built them. In Discovery Channel's *STRUCTURES*, you'll see how people have overcome challenges in design, planning, and construction to make bigger and better structures a reality.

STRUCTURES

Structures 4
At-a-Glance See how one structure stands tall despite the forces that work against it.

I Can't Take All This Tension 6
Q & A A Greek column explains how modern structures are able to face forces that were its downfall.

Bridging the Gaps 8
Scrapbook Newspaper accounts detail the best uses for three bridge styles.

No Pain, No Gain 10
Picture This Learn how architects, engineers, and builders solved problems to create world-famous structures.

Bigger and Better 12
Almanac Travel through time to watch the tallest skyscrapers rise. Compare bridges in the United States to those in other countries. And learn what went into the world's most famous five-sided structure.

Field of Dreams and Dollars 14
Timeline Trace the building of the Seattle Mariners' Safeco Field, to see how things add up.

Built by Design 16
Scientist's Notebook The Experience Music Project is unique in every way. Find out how the experts turned an architect's dream into a building.

Disappearing Act............ 18
Virtual Voyage Fly above the Chesapeake Bay Bridge-Tunnel Complex as crews repair the damage made by nature and time.

Three of a Kind.............. 20
Heroes John Roebling designed the Brooklyn Bridge, which would never have been finished without help from his family.

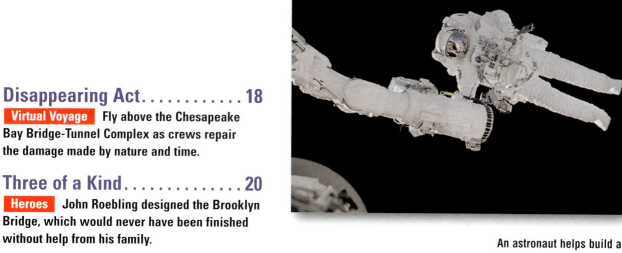

An astronaut helps build a structure in space. See page 22.

All Systems Going......................... 22
Amazing but True With half its construction carried out in space, the International Space Station is shaping up to be one of history's most complicated building projects.

Ancient Wonders of the World............... 24
Map Tour the world to learn about ancient structures that left reminders of great civilizations.

Reaching for the Sky....................... 26
Eyewitness Account See how workers finished the Empire State Building in record time.

Picking the Perfect Playground................... 28
Solve-It-Yourself Mystery Can kids design playground equipment that's fun and safe? You be the judge!

Brain Builders and Busters 30
Fun & Fantastic Check out the future of skyscrapers and bridges. Learn about a country that used one-tenth of its population to build the world's longest tunnel.

Final Project
Use Your School as a Tool.......... 32
Your World, Your Turn Build your own school. Use your math and art skills to create a floor plan and scale model.

AT-A-GLANCE Structures

The Washington Monument must always be the highest structure in Washington, D.C., according to federal law. At a height of 555 feet and 5⅛ inches (170 m), it is the world's tallest freestanding stone structure.

It's not easy being tall. There are universal laws that affect all upright structures—the laws of nature. Gravity is a natural force on Earth, pulling everything down toward the center of the planet. Structures also have to contend with weather and other forces, such as earthquakes, that push them from side to side.

A typical tall building's weight is held up by metal support columns. These columns allow the building to sway slightly in reaction to these forces. But the obelisk-shaped Washington Monument isn't typical. Except for an aluminum pyramid at its top, it is made entirely of granite and marble blocks. No columns support its 90,854 tons (82,441 tonnes).

By late 1997, the 115-year-old monument was ready for repair. Besides the usual wear and tear, the monument had long cracks running about three-quarters up one side. Rainwater seeping into the cracks caused even more damage.

During an 18-month project, workers built 37 miles (60 km) of scaffolding, sealed 500 feet (152 m) of cracks, cleaned 59,000 square feet (5,481 sq m) of wall space, and repaired 1,000 square feet (93 sq m) of chipped stone. To cover the biggest cracks and keep new ones from opening, workers used a soft mortar mix that moves with the monument when it sways.

Building experts at the monument and elsewhere use materials and techniques that allow buildings to handle the stresses of being on a planet where gravity is a central force—a force that has been trying to pull structures down since the first ones were built.

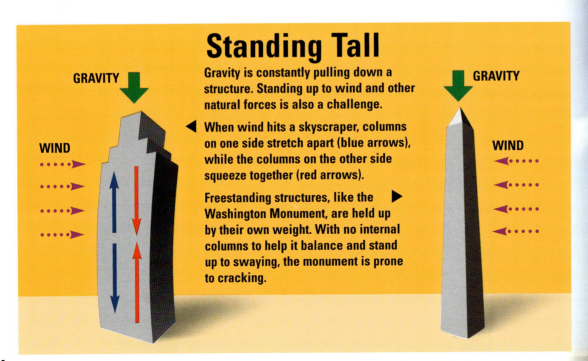

Standing Tall

Gravity is constantly pulling down a structure. Standing up to wind and other natural forces is also a challenge.

◀ When wind hits a skyscraper, columns on one side stretch apart (blue arrows), while the columns on the other side squeeze together (red arrows).

Freestanding structures, like the Washington Monument, are held up by their own weight. With no internal columns to help it balance and stand up to swaying, the monument is prone to cracking. ▶

The Washington Monument (right) remains a favorite tourist attraction in the nation's capital—even when surrounded by scaffolding (above).

Q & A

I Can't Take All This TENSION

Advice from a column about supporting big structures

Q: We're on the outskirts of a Greek city, talking with the last standing column from an ancient temple. Say, old-timer, got a minute?

A: How about a few hundred years? That's how long I've been standing around without anything to do.

Q: No offense, but I don't often think of a column as actually doing anything. Aren't you designed to be decorative?

A: Everyone thinks that! I know we're handsome. Still, some of us do more than stand around museums and libraries looking pretty. We're vertical supports that help hold buildings up. I myself used to hold up the temple that once stood here. Not alone, of course. There were probably a hundred of us.

Q: Quite a crowd!

A: Well, we had quite a big job to do—fighting all the forces trying to take the temple down.

Q: What forces?

A: Gravity, for starters. That's a biggie. Gravity tries to pull everything down. At the same time, huge slabs of concrete and marble—the things we were trying to hold up—were pushing down on us. So we had pulling forces, or tension, and pushing forces, or compression, going on. That added up to a VERY tiring thousand years!

Q: That does sound heavy-duty!

A: But that's not all. We were constantly being pushed and pulled horizontally too.

Q: By what?

A: Wind, rain, earthquakes—the kinds of forces you can see at work when tall, freestanding structures sway back and forth in a hurricane. All that wear and tear is what caused these cracks.

Q: You've been through a lot, haven't you?

A: Yep. It's a wonder I've stood it all these years. But you know, there's an upside to all that pushing and pulling. It helps buildings keep their balance.

Q: And that's important, right?

A: Well, yes, of course! All standing structures have to be built to balance. Otherwise they'd slump to the ground as soon as construction began, and what would be the point of that? Another thing that helps a structure balance is a bigger base.

Q: Like yours?

A: Right. Bigger bases give more structural support, making it easier for us to stand on our own. They also help distribute the compression from the weight above us. Spread it out.

Q: Still, with all the stress buildings have to take, it's a wonder any of them are still standing.

A: Thank engineers and architects for that. Good architects design buildings with an understanding of the forces the structure will face. Engineers build structures using the correct materials and methods. Figuring all that out isn't easy. Besides considering the weight of the structure itself, which can add up to millions of pounds, they have to factor in the weight of everything in or on the structure.

Q: Everything?

A: Right. People. Cars. Furniture. Paintings on the wall. Goldfish in the bowl. Meatballs in the spaghetti. And that's the part that gets tricky.

Q: Because some people don't eat meat?

A: No, because those things change. Some days hundreds of cars cross a bridge. Other times, like on holidays, thousands might cross. An empty apartment pushes down with less force than one full of furniture and a family.

Q: So architects and engineers have to plan for the heavier times and the lighter ones.

A: Correct. At festivals, there were thousands of people in my temple. Other times, not a soul.

Q: You must miss the old days, when columns held up the major buildings.

A: But columns are still holding up major buildings! You just don't notice them as much because they're likely to be inside the outer walls. Nowadays, columns are designed to take up less space.

Q: Oh. Any other differences?

A: They're made of metal—usually steel—the same materials that the horizontal beams are made of. Nothing's stronger than steel, which is really important in modern times.

Q: Why's that?

A: Your buildings have to work harder today than we ever did. They're so much taller, which means much more powerful winds blast them. You need materials that are strong enough to stand up to the challenge, as well as different building techniques and styles.

Q: So you approve of the changes in building design and construction?

A: Sure. The changes keep modern structures solid. Sometimes I miss the good old days. Toga parties. But I don't miss the tension.

Activity

WEIGHT AND SEE The way weight is distributed can affect the structure holding it up. Use modeling clay to make two large columns. Lay a piece of strong cardboard on top of them, creating a platform between them. Now test your columns' strength with a collection of books; spread them out so the weight is distributed. Also try wooden blocks stacked up in one or two places. Record your observations of what happens.

SCRAPBOOK
BRIDGING THE GAPS

Suspension, truss, and arch bridges aren't built for their looks. Engineers design a bridge based on what will work best in each construction setting, addressing the ever-present twin forces of tension and compression.

Suspension Bridge

Arch Bridge

Truss Bridge
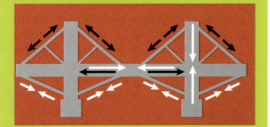

Black arrows show pulling (tension); white arrows show pushing (compression).

Note: All articles on these pages are fictitious.

The San Francisco Globe, May 27, 1937
MIGHTY BRIDGE SPANS GOLDEN GATE STRAIT
BY JOSEPH DALY

After four years of construction, the Golden Gate Bridge finally opens to pedestrians today. Vehicle traffic will be allowed on the bridge tomorrow.

This 4,200-foot-long (1,280-m) and 90-foot-wide (27-m) wonder is a suspension bridge, a design perfect for long stretches. Like most suspension bridges, the Golden Gate has twin steel cables anchored in concrete towers that hold up its roadway. Traffic and the structure itself cause compression, which presses the span downward, but the cables distribute the weight to the towers. As a result, each tower supports an astounding 61,500 tons (55,805 tonnes). Including the suspended structure, concrete anchorages, and approaches to the bridge, the total weight of the bridge is 894,500 tons (811,670 tonnes).

The cables themselves are an engineering marvel. Each of Golden Gate's cables are made of 27,572 steel wires twisted and bundled into 61 strands, which are further twisted. Individually, a cable measures 7,650 feet (2,332 m) long and more than 3 feet (.9 m) wide.

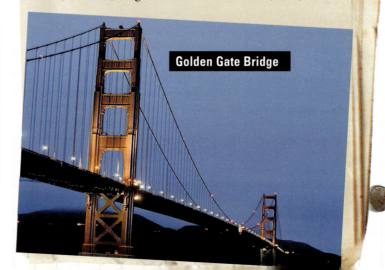

Golden Gate Bridge

Sydney Harbor Bridge

The Sydney Afternoon Record, March 19, 1932
Captain Arrested at Bridge Opening
BY LARRY FOX

Crossing one of the oldest bridge designs in the world with the latest design concepts and materials, officials today unveiled the Sydney Harbor Bridge. Yet a strange incident occurred at its opening. Just before New South Wales Premier John T. Lang cut the ceremonial ribbon, Captain Francis De Groot of the New Guard sliced it with his sword. He was quickly arrested, and the ribbon was retied so Lang could cut it properly. De Groot protested that he was caught up in the excitement of the moment.

The harbor was teeming with ships, waiting to pass underneath the widest, heaviest steel arch bridge in the world. Nearly one million people attended the opening of the bridge, which is 1,650 feet (503 m) long.

The Sydney Harbor Bridge is an arch design, which spreads out the forces of tension and compression so that no section of the bridge experiences too much weight. The sides of the arch squeeze together as traffic moves across the span. This tension spreads along the curves to the support columns on each end. The columns push back on the curves, preventing the ends of the arch from spreading apart.

Construction of the bridge took nine years. Everyone who witnessed its fully finished splendor today knew the same thing: This bridge will become an internationally recognized symbol of Sydney.

The Albany Whistle, March 16, 1888
Nation Mourns the Father of Iron Bridges
BY SALLY HARPER

Squire Whipple passed away yesterday. He was 84. A noted engineer and inventor, often called the "Father of Iron Bridges," Whipple attended Union College in Schenectady, New York, the first American college to develop an engineering program.

Whipple achieved national recognition in 1847 for publishing the first scientific and mathematic rules for building bridges. Before Whipple's book, wooden trusses were built using expert estimations.

Whipple's biggest achievement was the cast and wrought iron truss, which adds strength from a series of iron triangles that absorb the stress beneath the roadway. Generally, a beam bridge used for short spans is the most straightforward design. It consists of a single roadway secured by piers on each side. The weight of anything traveling over the beam creates compression, or pushes down on the beam. This weight causes tension on the underside of the beam, stretching it out. Too much weight without strengthening the beam would collapse the bridge.

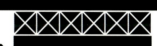

Beam bridge with Whipple truss

Activity

Number Decoder Use the chart at right to calculate how many of each animal it would take to equal the length or weight of each bridge on this page. As you go through the rest of the book, you may refer to this page to figure out more calculations, so you can understand and compare the properties of different structures.

SPECIES	AVERAGE WEIGHT	AVERAGE HEIGHT/LENGTH
African elephant (largest land mammal)	8 tons (7 tonnes)	13 feet tall (4 m)
Blue whale (largest mammal)	150 tons (136 tonnes)	110 feet long (34 m)
Golden retriever	75 pounds (34 kg)	24 inches long (61 cm)
Male giraffe	4,215 pounds (1,912 kg)	18 feet tall (5 m)

1 ton = 2,000 pounds

NO PAIN NO GAIN

A Tall Order

After construction began on the Petronas Towers in Kuala Lumpur, Malaysia, architect Cesar Pelli considered how to make the towers the tallest in the world. In 1998, the 1,483-foot-tall (452-m) towers grabbed the title of tallest skyscrapers. Pelli had found that increasing the number of floors would add too much weight. Yet he could add up to 243 feet (74 m) by extending the spires, or poles at the top of each tower. To make sure the buildings could withstand the added weight and forces without losing their balance, Pelli conducted a wind-tunnel test. It confirmed that the new design would hold up.

Some of the world's most famous structures caused major headaches for architects and engineers.

Taking Their Time

Constructing the Akashi Kaikyo Bridge between Kobe, Japan, and Awaji Island took 43 years of careful planning. To span the Akashi Strait, the suspension bridge had to be a record-breaking 12,825 feet (3,909 m) long. Completed in 1998, the 928-foot-high (283 m) bridge stands tall even during typhoon winds of 180 miles (290 km) per hour; it also remained standing after a strong earthquake. Its strength comes from a truss, which architects designed beneath the roadway. The truss also allows wind to pass through the bridge rather than slam into it. Each bridge tower has 20 tune-mass dampers (TMDs) to help reduce sway: When wind pushes the bridge in one direction, the TMDs push in the opposite direction, keeping the structure balanced.

Pinpointing Trouble Spots

The English Channel Tunnel—or Chunnel—stretches 31 miles (50 km) between Dover, England, and Calais, France. Before it was completed in 1994, architects, engineers, and workers had many concerns about digging 23 miles (37 km) under 150 feet (46 m) of water. Some 13,000 workers dug from both ends of the tunnel, aiming to connect in the middle. However, the distance between the starting points and the natural curvature of Earth made it difficult to predict the precise location of that meeting point. Engineers created mathematical models to correctly pinpoint it.

Of the Chunnel's three interconnecting tunnels, two are for trains and one can be used as an emergency exit. The extra tunnel added to the structure's cost of $21 million, but it has already saved lives. Thirty-one people escaped from a train fire one year after the Chunnel opened.

The train at left passes through a main tunnel, one of the Chunnel's three interconnecting tunnels (below).

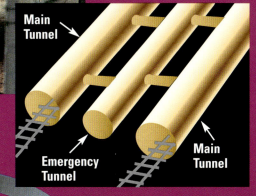

The Learning Curve

Architect Eero Saarinen designed the Gateway Arch so it would offer an amazing 30-mile (48 km) view. Made of stainless steel, the landmark stands 630 feet (192 m) tall next to the Mississippi River in St. Louis, Missouri. Saarinen realized that the arch's shape meant that the common elevator couldn't carry visitors to the curved top. To remedy this, a team of engineers created a tram with pivoting capsules—each holding five people—that travel up and down the arch's legs.

ALMANAC

BIGGER AND BETTER

At one point, each of the buildings below grabbed the title of tallest in the world. Height is commonly measured from the structure's base to the top of the highest architectural structure, not including antennae.

Tribune Building, New York, New York: This 10-story building is considered one of the first skyscrapers because of its use of steel and elevators.

World Building, New York, New York: Newspaper publisher Joseph Pulitzer made sure his building stood a mighty 309 feet (94 m) tall. The World Building was torn down in 1955.

Manhattan Life, New York, New York: This 348-foot-tall (106 m) insurance company building was demolished in 1930.

Park Row, New York, New York: This 30-floor building is 386 feet (118 m) tall, but lanterns on top technically make it 391 feet (119 m) tall.

Singer, New York, New York: Its 47 floors topped out at 612 feet (187 m). In 1968, the building was torn down.

| 1874 | 1890 | 1894 | 1899 | 1908 |

SPANNING THE GLOBE

Take a look at some of the longest bridges in the United States and see how they compare with bridges elsewhere in the world.

LONGEST SUSPENSION:			
In the U.S.	Verrazano-Narrows	New York, New York	4,260 feet (1,298 m)
Outside the U.S.	Akashi Kaikyo	Hyogo, Japan	6,529 feet (1,990 m)
LONGEST STEEL ARCH:			
In the U.S.	New River Gorge	Fayetteville, West Virginia	1,700 feet (518 m)
Outside the U.S.	Sydney Harbor	Sydney Harbor, Australia	1,670 feet (509 m)
LONGEST CONCRETE ARCH:			
In the U.S.	Natchez Trace	Franklin, Tennessee	582 feet (177 m)
Outside the U.S.	Wanxian	Wanxian, Sichuan Province, China	1,378 feet (420 m)

Chrysler, New York, New York: Architect Van Allen added 121 feet (37 m) to the 925-foot-tall (282 m), 77-story building by adding a spire.

World Trade Center, New York, New York: Both high-rises had 110 floors. Tower One measured 1,368 feet (417 m) high; Tower Two, 1,362 feet (415 m).

Petronas Towers, Kuala Lumpur, Malaysia: Both towers in these 88-floor structures reach 1,483 feet (452 m), making them the tallest structures in the world—so far.

Woolworth, New York, New York: This tower is 792 feet (241 m) tall with 57 floors. New engineering techniques kept it from swaying in the wind.

Empire State Building, New York, New York: After this 102-floor, 1,250-foot-tall (381 m) skyscraper was built, a 22-story TV antenna was added.

Sears Tower, Chicago, Illinois: At 1,450 feet (442 m), this high-rise is still tops in North America. From the skydeck, visitors can see four states.

| 1913 | 1930 | 1931 | 1972-1973 | 1974 | 1998 |

Sizing up the Pentagon

The Pentagon, headquarters of the United States Department of Defense, is a miniature city. As one of the world's largest office buildings, this five-sided structure in Arlington, Virginia, accommodates about twenty-three thousand employees. Here are a few statistics about this uniquely shaped structure.

- Construction took 16 months, costing about $49,600,000; it was completed on January 15, 1943.
- Approximately 680,000 tons of sand and gravel came from the nearby Potomac River to make concrete.
- The building is 77 feet and 3.5 inches (24 m) tall with seven floors.
- Each of the five outer walls measures 921 feet (281 m) long. That's about the length of 24 buses lined up end to end.
- Workers use 131 stairways, 19 escalators, and 13 elevators to reach offices that occupy 3,705,793 square feet (344,268 sq m).
- Phones connected by 100,000 miles (160,900 km) of telephone cable handle more than 200,000 calls every day.

Activity

ADD IT UP Using the numbers supplied about the Pentagon, how much office space can each worker have on average? Get a measuring tape and measure your principal's office. How does the size of this office compare to the size of an office in the Pentagon?

TIMELINE
Field of Dreams and DOLLARS

Talk about hitting one out of the ballpark! When officials in Seattle, Washington, decided to build a new stadium for their baseball team, they estimated its cost at $260 million. When construction ended, the Mariners' Safeco Field cost a whopping $517 million.

How did this stadium get to be so expensive? It costs a lot to build a stadium, especially if it's a state-of-the-art structure with restaurants, children's play and picnic areas, a museum, and a retractable rolling roof. The city paid more than 3,250 workers six days a week through an ever-expanding schedule. Below is a timetable of how the costly, mammoth structure developed.

⓫ 1996–1997: Construction Begins

Officials choose the field's site on **September 9, 1996**, but workers must clear the land. The ground is broken on **March 8, 1997**, and 23 days later, the first steel pile goes into the ground. The concrete foundation is poured soon after.

⓬ 1998: Tip-Top Shape

On **May 28, 1998**, on platforms 165 feet (50 m) above a nearby railroad, workers begin to construct the first of three roof panels. Each one will be mounted on 100-foot-tall (30-m) legs. The legs will allow a "raised roof" so batters have the chance to hit a home run out of the stadium. A shock absorber, which reduces the amount of sway caused by wind and earthquakes, is added to the top of each leg.

On **September 14, 1998**, the first roof panel is lifted 215 feet (66 m) over the field's steel frame. Construction of the other roof panels begins right away.

Safeco Field under construction on March 8, 1997 (above); on May 28, 1998 (right).

Safeco Field in Seattle, Washington

Budget Buster

Costs can add up. Contractors underestimated the amount of concrete needed for 10,000 yards (9,144 m).

Original budget for concrete: **$38 million**

Actual cost: **more than $60 million**

❸ 1998–1999: From the Field Up

By **August of 1998,** construction of the seating platform is complete, and a state-of-the-art playing-field drainage system is finished soon after.

In **March of 1999,** workers install the scoreboard with high-definition video images. The roof won't be ready to roll until June, but by **April 28, 1999,** two 800-foot-long (244-m) runways with steel wheels are built. Powered by ninety-six motors, the runways allow the roof to glide over the playing field.

Turf installation begins at the end of **May 1999,** but it will take at least seven weeks for the grass to sprout.

❹ 1999: Play Ball

On **July 15, 1999,** the Seattle Mariners play their first game in the stadium against the San Diego Padres for a full house of 47,155 fans. A patio with a batting cage and barbecue pits attracts thousands of fans. People can walk the entire ballpark on the main concourse without missing a moment of the game. People walking along one side of the stadium are able to peek inside through glass-covered openings.

Structural problems? There were a few. The biggest was that the roof didn't close completely during the second game and rain leaked onto the field. "We hit a home run," said Mariners Vice President Bob Alyward, referring to the new stadium. "But it wasn't quite a grand slam."

Safeco Field Stats

- Platforms cost $2 million.
- Shock absorbers total $70,000.
- The three roof panels contain enough steel to build a 55-story skyscraper.
- The longest roof panel is 655 feet (200 m), the length of 17 buses parked end-to-end.
- More than 600,000 bricks went into the facade, or the outside of the structure.
- The drainage system can handle up to 130,000 gallons (491,920 l) of water, making a wet field ready for play in 45 minutes.
- Inside the playing field are 150 miles (241 km) of electrical wiring and 55 miles (88 km) of heating pipe that warm the ground.

Activity

OVER YOUR HEAD The roof over Safeco Field is similar to a tent: It depends on balance. Using a piece of cardboard as a base, construct a tent with a paper towel, Popsicle sticks (to use as a frame), thread (to hold the sides down), and 4 inches of tape. What other kinds of structures are held up by balanced forces?

SCIENTIST'S NOTEBOOK

BUILT by DESIGN

What does a modern museum have to do with an electric guitar? In this case, everything.

It all started when Microsoft cofounder Paul Allen commissioned an architect to build a rock 'n' roll museum in Seattle, Washington. His instructions: "Make it swoopy. Really push it. Take it to the moon."

Architect Frank Gehry did just that when he designed the Experience Music Project, which opened in June 2000. The building is a showcase for rock 'n' roll, so Gehry looked to legendary musician Jimi Hendrix's electric guitars for inspiration. The instruments' curving heads and rich colors led him to create a structure without a single right angle on the entire exterior.

Allen's request was a perfect example of a human need addressed by science and technology. He knew his museum would have to be special to separate it from the Rock and Roll Hall of Fame and Museum in Cleveland, Ohio. He and his staff could handle finding the museum's unique contents, but Gehry and his team would have to break new ground with the building's design.

REALITY CHECK

The nontraditional building posed many riddles for the building contractors. Huge windows required custom fitting. The city's monorail would run through the building, so it had to provide plenty of space and withstand the pressure of the train's speed. Yet the biggest construction concern was choosing the kind of architectural support system that could hold up the 140,000-square-foot (13,006-sq-m) structure.

Gehry and his team of architects and engineers considered many frame styles, finally determining that the most durable construction would be a skeleton of steel-plated beams shaped like ribs in the human body.

Engineers created special machines to gently shape each identical beam. They made more than fifty, each one weighing between 5,000 and 10,000 pounds (2,268 and 4,536 kg).

Then the design team took over. Using the computer program CATIA (Computer-Aided Three-dimensional International Applications), the designers pressed the tip of a drill-type tool every half-inch along the surface of a miniature model of the museum. Each point was digitized into an electronic model that allowed the designers to "get inside" the structure before they created the real thing. CATIA allowed the designers to pinpoint where each beam should go, to create a balanced structure.

Jimi Hendrix's guitars inspired the design of the Experience Music Project.

A Seattle landmark, the Space Needle rises above the city's unusual new structure.

Under construction: the Experience Music Project in Seattle, Washington (left); its entrance is shown below.

SOLVING MORE PUZZLES

Once the building's frame was in place, workers began pouring concrete. The museum's unique shapes began to take form. Finally they created the building's second skin—a stainless steel exterior, made from the thickest steel available. "You can only torture metal so much," said Paul Zumwalt, director of design and construction. "The physics of metal just don't permit it. We eventually had to break the curves down into smaller sections to get the metal to behave."

The architecture of the Experience Music Project is one of a kind. "It's rock 'n' roll architecture," says Allen. "It's expressive and fresh and exciting. You're not going to see too many 120-foot-high (37-m) purple metallic structures."

Project manager Carolyn Bird sees other benefits, too. She believes that what engineers learned during the building of this museum will improve the way all buildings are constructed. "It has gone steps beyond what a conventional building would look like," she says. "We've shown it can be done."

Activity

BEND IT BACK Some materials can be bent without breaking. Find a twig and watch closely as you bend it slowly until it breaks. What happens to the top and bottom surfaces? Collect other bendable materials and see how they respond to bending. What do the materials that break have in common? What would make these materials sturdier?

VIRTUAL VOYAGE
DISAPPEARING ACT

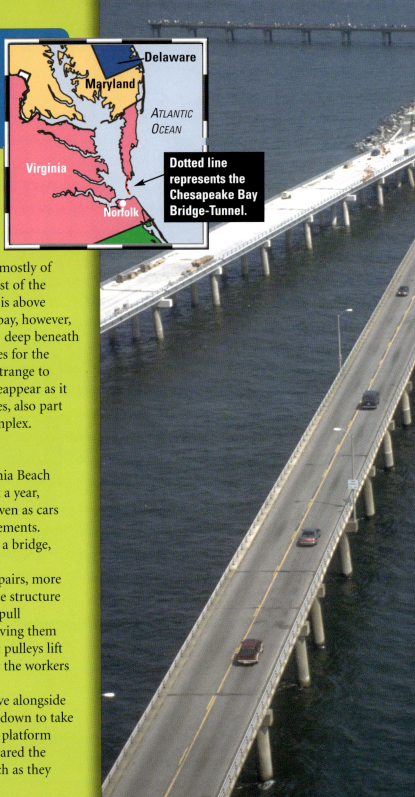

The Chesapeake Bay Bridge-Tunnel connects Virginia Beach/Norfolk to Virginia's Eastern Shore. This curved marvel stretches 17.6 miles (28 km). No wonder repairing it is an almost constant activity. Let's go in for a closer look.

Dotted line represents the Chesapeake Bay Bridge-Tunnel.

The sun is shining warmly for the first time today, and you—a herring gull—are perched on top of one of the bridge's 498 light poles. As the traffic begins to get heavier, you swoop down from the pole to look around.

It takes a while to fly across the world's largest bridge-tunnel complex, which consists mostly of low-level bridges with a two-lane highway. Most of the complex, supported by more than 5,000 piers, is above water. Because a lot of ships pass through the bay, however, two-mile-long tunnels lie about 40 feet (12 m) deep beneath the bay. Four artificial islands provide entrances for the roadway to connect to the tunnels. It's pretty strange to see the road disappear into the bay and then reappear as it comes out of the tunnel. Two high-level bridges, also part of the crossing, are at the north end of the complex.

A Bird's-Eye View

Just for fun, you decide to start at the Virginia Beach end and fly north to the other side. For almost a year, many workers have been painting the bridge even as cars drive past. The paint is a barrier against the elements. Corrosion, the formation of rust that weakens a bridge, is a big problem.

Today, like most days during the bridge's repairs, more than thirty-five workers arrive to make sure the structure stays in working order. You watch as tugboats pull equipment-filled barges through the water, leaving them in strategic spots alongside the bridge. Electric pulleys lift platforms on the barges into place. They allow the workers to get close enough to paint the span.

Rigged on trolley wheels, the platforms move alongside the bridge as crews paint each section. You fly down to take a closer look. There are pulleys that lower one platform back down to the barge. Once the crew has cleared the supports in the middle of the bridge, you watch as they

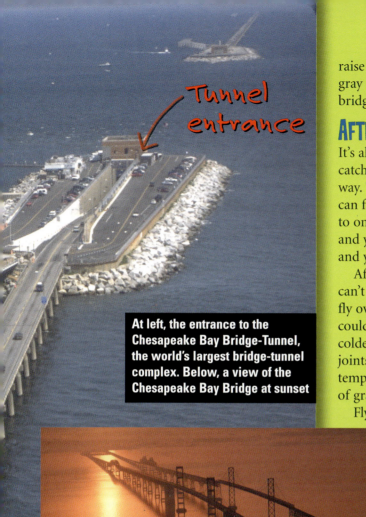

At left, the entrance to the Chesapeake Bay Bridge-Tunnel, the world's largest bridge-tunnel complex. Below, a view of the Chesapeake Bay Bridge at sunset

raise the barge back up into position again. You still see some gray spots on the bridge, but each day, more sections of the bridge are painted blue, to match the color of the sky.

AFTERNOON ACTION

It's almost lunchtime and you're getting hungry. You'd like to catch a fish you see underwater, but there's something in your way. The crew has covered the work area so that no materials can fall into the bay and contaminate the water. You fly over to one of the four artificial islands. There you can rest in trees, and you like to watch all the people who come to observe you and your bird buddies.

After lunch, you're feeling a little restless. You know painting can't be all that's happening at the bridge. And you're right. You fly over to another crew repairing dents in a bridge rail. Dents could turn into more serious cracks when the weather gets colder. Hmm. Another group of workers is cleaning expansion joints, which allow a bridge to safely contract or expand when temperatures change. The workers have to keep these joints free of grass and other debris so the joints can function properly.

Flying over several miles of the low-level trestle—the part of the beam bridge with slanting supports—you see a crew in front of a huge cement truck with a continuously rolling cylinder. It's a mixer, and it keeps turning so the concrete it holds won't dry and harden before the workers can use it to patch the concrete trestle.

Feel that chill? The weather is starting to get cold. Normally ice isn't a problem workers have to handle on this roadway. But the last time the road was icy, you watched six trucks spread sand down the four lanes so cars wouldn't slip.

Speaking of winter, it's time for you to fly farther south. You'll visit next year to see what else has been fixed at the bridge-tunnel since you were away.

Activity

NAILING RUST DOWN Take a look at what can happen to metal in a week's time under constantly wet conditions. Get a sturdy file from a hardware store and file four nongalvanized nails until they are roughed up. Leave one unpainted, paint another, cover another with baby oil, and put antirust spray on another. After letting them sit for three hours, place all four nails in separate, labeled plastic cups of water. Once a day for a week, use a tweezer to take each out. Examine each and write notes on your findings. Which rusts the fastest? Can you explain your results?

19

HEROES
Three of a Kind
A family's tragedy—and triumph—built the Brooklyn Bridge.

In the early days of bridge building, engineers were still developing safe tools and techniques. As a result, workers commonly had deadly accidents. Twenty-seven men died while building the Brooklyn Bridge in New York City, and one of them was chief engineer John Augustus Roebling. His son, Washington, and Washington's wife, Emily, took over the challenge. All three are heroes not just because of the risks they took, but also for the contributions they made to a historical landmark.

John Roebling

Working Toward a Goal

John Roebling found a way to twist steel wire into thick suspension cables in 1841. These bunched wires were strong enough to support heavier loads and longer spans than ever before. Seeing he had a product that would be in great demand, the German immigrant opened a New Jersey factory to offer his cables to other builders.

Roebling's invention excited many in the building industry because it was sure to improve upon the design and safety of bridges around the world. Add to this the experience Roebling picked up managing the construction of other bridges, including the Cincinnati-Covington Bridge in Ohio, and you have a very capable leader. The building community realized they had found the best man to tackle New York's biggest architectural challenge: a bridge that would connect the island of Manhattan with Brooklyn, a borough across the East River.

In 1869, state officials accepted Roebling's design for a steel-cabled bridge with stone piers. He decided the Brooklyn Bridge would be the first to use steel cable wire, and it would stretch 1,595 feet (486 m)—50 percent longer than any bridge at that time.

Roebling's vision was revolutionary, but fellow engineers grumbled that the job would be a tough one—if it worked at all. English engineer Robert Stephenson, who had the challenging task of building a bridge over the Menai Strait in North Wales, wrote to Roebling, "If your bridge succeeds, then mine have been magnificent blunders."

The state set aside $15,100,000 to build the bridge. Knowing this would be his masterpiece, Roebling decided to guide the project as chief engineer.

Death Before Glory

Standing on a dock before construction began, Roebling was taking compass readings to determine the location of the Brooklyn tower. A ferry slammed into the dock, catching one of Roebling's feet between two steel piles. His injury was so bad that doctors had to amputate some of his toes.

Three weeks later, Roebling died of tetanus, a disease caused by his infected wounds. It was July 1869, and work on the bridge was not expected to be finished for at least another ten years.

Knowing his father's devotion to the project, Washington Roebling took over as chief engineer. Construction began on January 3, 1870.

Like his father, Washington preferred to direct workers on-site, rather than from his office. He often visited the caissons—watertight underwater structures in which workers were constructing the tower foundations. The caisson on the Manhattan side was 78 feet (24 m) below sea level; on the Brooklyn side, the caisson was 44 feet (13 m) deep.

Traveling quickly from a high-pressure underwater environment to the lower pressure above ground takes a toll on the body. Nitrogen bubbles began to form in Washington's blood. This dangerous condition is commonly called the bends by scuba divers.

By 1872, Washington was paralyzed, partially blind, and deaf. He could not leave his bed. Yet from his home in Brooklyn, he could watch the bridge's construction through a telescope.

Washington supervised every step with help from his wife, Emily. She visited the building site daily to relay her husband's instructions. At times, Washington worried he couldn't finish the job but, as he wrote, "I had a strong tower to lean upon . . . a woman of infinite tact and wisest counsel."

Washington's health was so bad at one point that doctors believed he was near death. Still, Washington didn't want to turn the job over to another engineer. Instead, he explained to Emily every detail necessary to finish the bridge, so she could carry on his work. Every day for four months, Roebling dictated instructions to her. She would take notes until he collapsed into a deep sleep.

Washington Roebling

The Brooklyn Bridge under construction in the 1870s

© Collection of The New-York Historical Society

Emily Roebling

Dream Come True

In the end, Washington lived well past the opening of the Brooklyn Bridge to traffic on May 24, 1883. Festivities on that day included fireworks, a boat procession, a parade in Manhattan, and a visit from President Chester A. Arthur and New York's governor, Grover Cleveland. One shopkeeper put a sign in his window that read, "Egypt had her Pyramid, Athens her Acropolis . . . so Brooklyn has her bridge."

Activity

STRENGTH IN NUMBERS Create a competition with your classmates to see who can build the strongest bridge. Separate into four groups to plan and construct a bridge made of plastic straws, tape, paper clips, and yarn. Test the bridges by gently rolling the following items over the roadway (in this order): a small toy car, a tennis ball, and a baseball. Do all of the structures hold up under all of the different weights? Why or why not? Make suggestions that will improve structures that didn't stay up. Then adjust the bridges and test them again.

All Systems Going

If you think building a structure on Earth is tough, imagine the challenges of building the International Space Station—while it's in space.

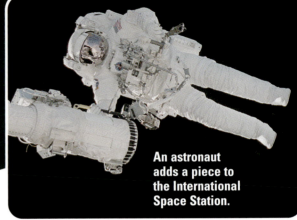

An astronaut adds a piece to the International Space Station.

The International Space Station (ISS) will allow astronauts to live in space for long periods of time, collecting data on their surroundings and the effects on the body of extended stays in space. Some officials compare its construction to building a giant Lego system in space.

An international team of architects and engineers worked together during the design process. Their task: to devise solutions to the problem of building such an enormous structure in space. In the first construction phase in 1998, Americans and Russians linked modules, or space capsules. Today, sixteen countries are collaborating to complete the $60 billion structure by 2006. It's the most difficult, dangerous, and expensive construction project in history. "We're starting to embark on a set of activities that are probably as complex as anything that we've ever done in the space business," says National Aeronautics and Space Administration (NASA) official Ron Dittemore.

Worth the Weight

Unlike most modern buildings on Earth, most parts of the ISS are a combination of metals, including aluminum. But weight isn't a construction problem. Much of the ISS will be constructed in space because many finished pieces would be too heavy to transport from Earth.

The completed station will be the largest artificial structure in space, weighing about 1 million pounds (453,600 kg) and covering an area as large as two football fields. Separate shuttle trips will deliver the main modules, including a laboratory and living quarters. The more than one hundred ISS pieces constructed on Earth are carefully made to fit those already in space. After all, it's not easy to return a piece if it's the wrong size.

By October 2000, construction was about half finished. Astronauts had attached two cargo cranes, a glass cockpit, and the first section of the station's "spine," a 356-foot-long (109-m) frame that holds communications instruments.

The ISS in September 2000

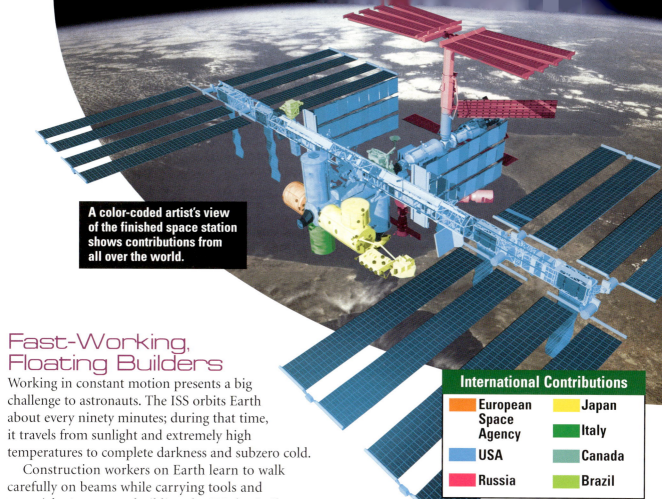

A color-coded artist's view of the finished space station shows contributions from all over the world.

Fast-Working, Floating Builders

Working in constant motion presents a big challenge to astronauts. The ISS orbits Earth about every ninety minutes; during that time, it travels from sunlight and extremely high temperatures to complete darkness and subzero cold.

Construction workers on Earth learn to walk carefully on beams while carrying tools and materials. Astronauts building the ISS do similar work, but they are balancing in a weightless environment more than 200 miles (322 km) above Earth. While astronauts don't have to worry about the weight of the pieces they are carrying, they must use great care to prevent materials from floating away.

Astronauts also have to rely on the shuttle's power supply, so they must often work quickly to make sure they'll have enough power to return to Earth. NASA officials estimate that astronauts will take at least 140 space walks to complete all construction.

In September 2000, American astronaut Edward Lu and Russian cosmonaut Yuri Malenchenko attempted one of the ISS's boldest building adventures. They moved farther away from a space shuttle than any other astronaut had done before. A robotic arm on the shuttle transported the men 40 feet (12 m) away to the station. When they reached the ISS, Lu and Malenchenko—wearing 300-pound (136 kg) spacesuits—climbed more than 60 feet (18 m) to attach and test an instrument that will measure Earth's magnetic field. They floated as they worked for more than six hours.

International Contributions
- European Space Agency
- USA
- Russia
- Japan
- Italy
- Canada
- Brazil

All the challenging work is adding up. At the end of 2001, the station measured 1,800 square feet (167 sq m) long, the size of the average three-bedroom house.

Activity

UNDERWATER WORLD Building underwater is a lot like building in space. Objects float away, and materials must handle a special environment. Using Popsicle sticks and coated wire, build a five-sided miniature house in your bathtub or in a large bucket of water. When you are done, write a step-by-step account of what you did, including any problems you had. How do you think this activity would have been different if you were underneath the sea?

Ancient Wonders of the World

For many centuries, engineers and architects have been applying creativity to their knowledge of physics and nature to solve problems and create monuments. These ancient wonders are the proof.

❶ COLLINSVILLE, ILLINOIS Between AD 700 and 1500, the Woodland Indians built a series of mounds out of 50 million cubic feet (1.4 million cubic m) of earth. The largest is Monks Mound; it covers 14 acres (6 hectares) and rises nearly 100 feet (30 m). Scientists believe tribe members carried the dirt in woven baskets and arranged the mounds to create a base for a 50-foot-high (15-m) wooden temple.

❷ MEXICO CITY, MEXICO Aztec engineers used their knowledge of science: With help from ramps that were continuously made higher, workers constructed the Great Temple during the 14th and 15th centuries. The stone temple stood about 100 feet (30 m) high, and its shrines on the top were dedicated to the gods of war and rain.

❸ ALEXANDRIA, EGYPT The Lighthouse of Pharos was built in the third century BC. It was about 400 feet (122 m) tall. Thanks to the scientific expertise of the designers, workers could rely on horse-drawn carts to haul marble up a spiral ramp inside a section 200 feet (61 m) high. It stood for more than a thousand years, until an earthquake destroyed it.

❹ GIZA, EGYPT The Egyptian Pharaoh Khufu ordered the construction of the "most amazing tomb ever built" in about 2600 BC. Known today as the Great Pyramid of Giza, it contains about 2,300,000 blocks of stone, each weighing 22 tons (20 tonnes). Archaeologists believe the engineers knew that a slippery, mud-soaked ramp would help hundreds of workers haul stone blocks 481 feet (147 m) high.

❺ **SEGOVIA, SPAIN** Between 312 BC and AD 226, the Romans built eleven aqueducts—structures for transporting water from one place to another—in Segovia and throughout the Roman Empire. The aqueducts could move water up to 57 miles (92 km). The Roman engineers knew that gravity would do all the work: Each aqueduct had a slope to let the water flow naturally over valleys and other low areas.

❼ **NEAR XI'AN, CHINA** Piling up dirt and stone into narrowing steps, an army of Chinese workers built the White Pyramid around 700 AD as a tomb for the Empress Wu Hou. At about 1,000 feet (305 m) tall, it is the world's tallest pyramid. No one is sure how it was built. It was so well hidden that explorers didn't discover it until 1947, when a U.S. pilot photographed what he thought was a huge dirt mound.

❽ **CHINA** About 220 BC, the Chinese began building the Great Wall. Designed to protect against invaders, it took workers a thousand years to complete. At first they used bricks made from wooden frames packed with dirt; later, engineers knew concrete bricks would be stronger. The Great Wall is 30 feet (9 m) high in some places, and, at 4,500 miles (7,241 km) long, it is the longest structure ever built.

❻ **ROME, ITALY** The Romans built the Colosseum out of stone and concrete to accommodate the 50,000 spectators who came to see fighting matches and other entertainment. Completed around AD 70, this huge amphitheater took ten years to build. Before then, theaters were built into hillsides, which provided support. The freestanding Colosseum stood 157 feet (48 m) high. Engineers determined that a series of arches around the building would distribute its weight.

Background photo: Colosseum in Rome, Italy

Activity

BUILDING WITH THE BASICS Create a pyramid, with a triangle as its base, from craft sticks and glue. Is it sturdy? What might it be able to hold? If the pyramid pieces were as tall as you, what kinds of things could you make from this structure? What would happen if you attached several together to make one structure? Can you build different things with more than one pyramid?

25

Reaching for the Sky
New York City, 1929-1931

A project this massive and difficult had never been attempted before. From gut-wrenching heights, steelworkers, called "sky boys," would dangle from thin steel beams more than 1,000 feet (305 m) in the air above concrete walkways. It was all part of building the world's tallest skyscraper: the Empire State Building.

Two years after construction began, a reporter described the workers' remarkable accomplishment: "Like little spiders they toiled, spinning a fabric of steel against the sky. Crawling, climbing, swinging, stooping—weaving a web that was to stretch further heavenward than the ancient tower of Babel."

A "sky boy" works on the Empire State Building. At right, the Chrysler Building towers over New York City.

Getting Down to Business

Two goals faced the architects and engineers of the Empire State Building: Build the tallest skyscraper in the world and do it in a year and a half. Why so quickly? Pride and competition were at stake. Walter Chrysler, founder of the Chrysler Corporation, had just begun to construct the "world's tallest building." A group of business leaders, which included the head of General Motors, formed a corporation to top him.

Amazingly, construction workers finished building the 1,250-foot-tall (381-m), 102-story structure in one year and forty-five days—record time to this day. Its record-breaking size knocked the Chrysler Building to second place.

The tight deadline demanded extensive planning and teamwork, with more than a thousand workers on site, seven days a week. The architects revised the plans seventeen times. "I doubt that there was ever a more harmonious combination than that which existed between owners, architects, and builders," boasted Paul Starrett, one of the builders. "We were in constant consultation with both of the others; all details of the building were gone over in advance and decided upon before incorporation in the plans."

Up, Up, and Away

Construction began in February 1930. Fascinated by the activity, New Yorkers watched as workers dug a hole 55 feet (17 m) deep and laid the foundation.

Times were hard. The country was experiencing the Great Depression, when many people had lost their jobs and savings, and many businesses were failing. The project offered people hope and employment. Those fortunate enough to be working were eager to see the building process run smoothly and quickly.

According to one of the skyscraper's architects, Richmond Shreve, "[Sometimes] things clicked with such precision that once we erected fourteen-and-a-half floors in ten working days—steel, concrete, stone and all. We always thought of it as a parade in which each marcher kept pace, and the parade marched out of the top of the building, still in perfect step. Sometimes we thought of it as a great assembly line—only, the assembly line did the moving; the finished product stayed in place."

The work was difficult, but there were no major problems. The engineers made use of the most recent advances in design. They decided that walls did not have to support the whole structure. Instead, a skeleton of 60,000 tons (54,444 tonnes) of custom-made steel beams would be strong enough to carry the weight of the building and insure that the forces of tension and compression stayed in balance. Large cranes helped the steelworkers move the beams into place, where they were fastened together with steel bolts.

The Empire State Building under construction in 1930

The Empire State Building today

The Gang's All Here

To get the job done quickly, building phases overlapped. One group, or gang, of workers built brick walls, while another put wiring and plumbing systems in place. As gangs finished framing a floor and moved up to the next, cement and stone workers enclosed that floor, and another gang of workers moved in to create window frames and interior walls. The building grew, and cranes raised the steel support beams ever higher.

The construction setting of a busy city made it impossible to keep materials on-site until the day they were needed. Steel was formed and cut to fit at mills in Pittsburgh, Pennsylvania. Trucks brought it to Bayonne, New Jersey, where it traveled by barge to the city. Trucks carried the steel beams to the site; each beam was marked to show its exact location in the building.

Nine electric-powered derricks—machines connected to beams to move heavy objects—lifted the girders into position, where workers bolted and riveted them together. As the building rose, miniature railroad cars carried up the supplies.

A well-defined schedule showed what would be happening at every minute of the day. Each morning, builders were told which gang of workers could expect a car filled with tools and materials. Work was so systematic that builders rarely had to ask for tools or equipment.

When the Empire State Building opened on May 1, 1931, everyone celebrated the strength and hopefulness of the workers who had constructed it and the intelligence and skills of its designers, builders, and architects.

Activity

Up with Paper Build the tallest structure possible using four sheets of 8 ½" x 11" construction paper. Do not use adhesives or scissors. You may tear or fold the paper into any size. Measure your tower. How much higher do you think you could make the structure if you had ten pieces of 1-inch long tape? Try it. Think of five unique ways you could make this structure stronger, and try at least two of them. Test your ideas by gently shaking the surface the structure stands on or blowing a gentle fan in its direction.

SOLVE-IT-YOURSELF MYSTERY

Picking the Perfect

"... and the winning team will have their names embedded in stone as the junior engineers of Keystone County Middle School's playground," announced Principal Skye Scraper to the cheering auditorium. "I'm sure it will be very difficult for our judges to decide which team has the best new playground structure. Remember, safety is the main concern in a playground, so we've asked judges to rank entries based on two primary criteria: 1) the safest design, and 2) the safest choice of materials used in that structure. The entry with the highest marks in both categories will be used to make the blueprint for the centerpiece of our new playground!

"Students, keep some basic principles in mind as you think about what kind of structure to create. The materials you choose should be able to withstand tension, or forces that pull, and compression, or forces that push."

"This is so exciting," Eddie Fiss said to his teammates. "I know we'll have a sure chance of winning this contest. All we have to do is think of the coolest playground structure we've ever seen and go from there!"

"What about some kind of open dome?" suggested Guy Telamon. "It's completely balanced, so it won't tip over. If we build a big one—say, 20 or 30 feet (6 or 9 m) in diameter—kids could climb all over the frame. Then we could hang swings on the inside."

"Yeah, but then the frame should be made out of something strong, like steel, so that it could support the swings' weight that pulls down," said Eddie. "That's especially important for a tall guy like me who loves to swing as high as humanly possible!"

"Actually, I think an aluminum frame would work better," said Guy. "It's strong, too, and can hold up to pulling and pushing. Aluminum can bend easier

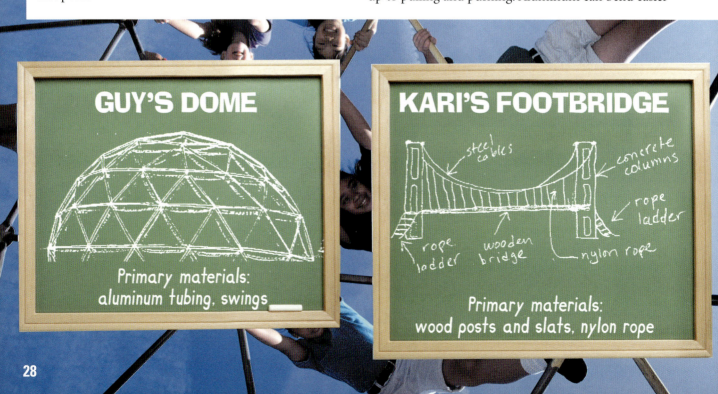

GUY'S DOME
Primary materials: aluminum tubing, swings

KARI'S FOOTBRIDGE
steel cables, concrete columns, rope ladder, wooden bridge, nylon rope
Primary materials: wood posts and slats, nylon rope

28

Playground

than steel, so it would be a better material to make a round structure."

"That sounds pretty cool," agreed Kari Bedrock. "But I think it would be fun if we could have a long footbridge. It could be the kind that has wooden slats strung together with nylon rope, and it will flex, or bend, when kids run across it. Each end could have two wooden posts to support the bridge with ladder rungs between them for kids to climb. I think a good safety feature would be to put lots of sand under the bridge instead of using asphalt or concrete. That stuff hurts when you fall on it! Sand would make a great cushion if someone fell, plus it would be like being at the beach!"

"Hmm," mused Eddie, taking in his friends' ideas. "I was sort of thinking about a pyramid—something equally balanced on all sides. The basic framework could be made of four steel beams joined at the top, so it would be super strong. There could be monkey bars that span across the center of the pyramid, and the sides could have nylon nets. That way, there'd be plenty of places for kids to climb."

"Well," said Kari, "it looks like we've got some great ideas to work with. Let's draw our plans and share them with Mr. Archer, the science teacher. Maybe he can tell us which plan will be the best one to submit to the school contest."

Three weeks later, Eddie, Kari, and Guy circled around Mr. Archer as he reviewed their plans. "You've all come up with some solid structural blueprints, but only one of these will have a real chance to win the contest."

Which structure did Mr. Archer think was the best one and why? Look at each plan and read the clues below to determine the best contest entry.

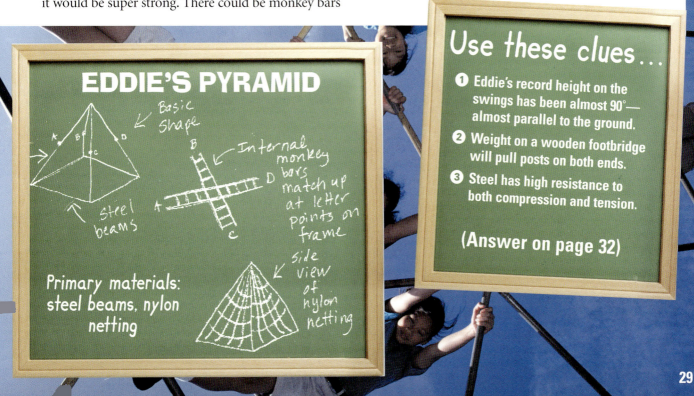

Use these clues...

1. Eddie's record height on the swings has been almost 90°—almost parallel to the ground.
2. Weight on a wooden footbridge will pull posts on both ends.
3. Steel has high resistance to both compression and tension.

(Answer on page 32)

BRAIN BUILDERS and BUSTERS

NOT SO FAST

Too bad, Petronas Towers! Just when you think you're tops in height, China plans to build a skyscraper to topple your title. Shanghai World Financial Center will stand tall at 1,509 feet (460 m) with at least 95 floors. It won't hold the title for long, though. South Korea's 1,516-foot-tall (462 m) Suyong Bay Tower should be completed in the near future. Chinese engineers say they may add to their building, but they won't say how much, because they don't want to let Suyong Bay builders in on their secret.

The Petronas Towers in Kuala Lumpur, Malaysia, are the tallest buildings in the world—so far!

What did one wall say to another?
I'll meet you at the corner!

What kind of building is the tallest in the world?
A library—it has the most stories.

How many 6-inch by 6-inch (15-cm by 15-cm) blocks would it take to complete a building that's 20 feet (6 m) long and 20 feet high on all four sides?
Only one. The last one you put in completes it!

The Sky's Not the Limit

Some architects dream of buildings that stretch far beyond any sky-high structure you've ever seen. Almost ready to sprout up in Hong Kong is the 1,883-foot-long (574 m) Kowloon MTR Tower. Not to be outdone, the Maharishi Mahesh Yogi has his own record-breaking building plans. He hopes to build a learning center in India that reaches 2,222 feet (677 m) into the sky.

Are these ideas possible? Sure, when the technology becomes available. Current technology won't allow buildings higher than 150 stories, so how can engineers overcome this limitation?

- Design elevators that require less space in buildings.
- Develop lighter, stronger building materials to handle the weight of additional floors.
- Increase wind resistance. Tall structures built to sway in wind could make people seasick and cause windows to pop out.

Style and Attitude

Most concrete and steel bridges don't last longer than twenty years without requiring major repairs. Technicians at the University of San Diego have been testing a bridge material mix that includes glass. Wear-and-tear results have been so promising that engineers are planning to build a 450-foot (137 m), four-lane "glass bridge" across a large highway in San Diego, California.

Glass- and carbon-fiber-reinforced plastic bridge decks are replacing some highways in Ohio and California. More expensive than concrete, this material is about one-fifth lighter and more resistant to earthquakes.

Researchers are also searching for ways that bridges can monitor themselves. If a bridge gets cracked in an earthquake, it would send a computer message that pinpoints the problem to a repair crew. This advance would take a while to put into action, so "complaining" bridges won't be around any time soon.

EVERYBODY DIGS IT

To build Japan's Seikan railroad tunnel, the longest in the world, it took twenty-five years and more than 13,800,000 workers. That's more than one-tenth of the country's population!

Seikan railroad tunnel

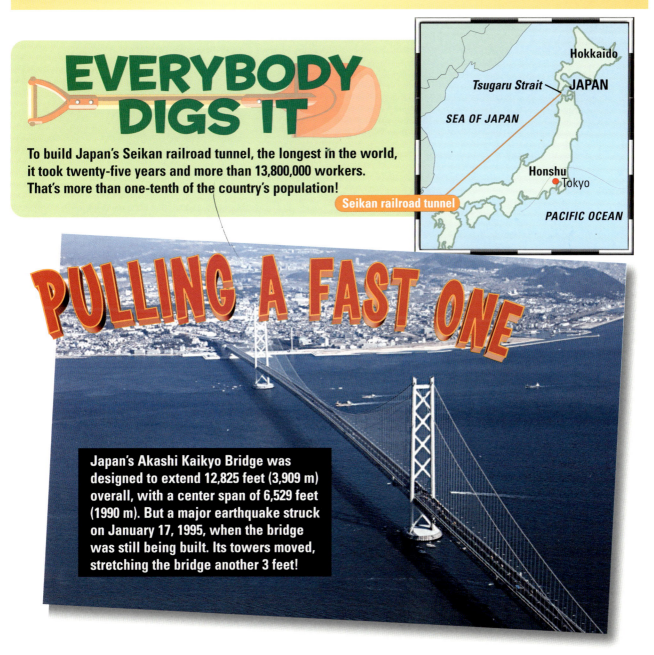

PULLING A FAST ONE

Japan's Akashi Kaikyo Bridge was designed to extend 12,825 feet (3,909 m) overall, with a center span of 6,529 feet (1990 m). But a major earthquake struck on January 17, 1995, when the bridge was still being built. Its towers moved, stretching the bridge another 3 feet!

Use Your School as a Tool

It may be hard to imagine, but building a large structure like your school required a team of architects, engineers, and construction workers to work months, maybe even years. After their design was approved but before they could start building, the architects probably built a scale model. Now you can create a scale model and floor plan of the same structure.

First, you'll need to measure the walls, windows, doors, ceilings, and all structural fixtures, such as columns, in your school. Use this information to create the floor plan. It does not have to be as technical as the actual blueprints used to build your school. Talk to an architect or engineer to help you decide exactly what information you'll need and how you can present it. If your school is fairly new, you may try to contact the firm that constructed it.

The next step is to create the scale model. Decide how big your model will be and modify your measurements so that it will replicate your school's design. Your class can decide which materials would look best, but you may consider using clay, cardboard, wire, or wooden blocks, and painting the finished product to match your school's colors.

ANSWER Solve-It-Yourself Mystery, pages 28–29

Mr. Archer picked Eddie's pyramid as the team's best choice. The triangular shape of each side of the pyramid structure and the four main steel beams will provide a solid, supportive framework for internal monkey bars and external climbing nets.

Guy's dome, like Eddie's pyramid, is also well balanced. His mistake was including swings inside the dome. Kids need plenty of clearance when swinging. Installing swings on the inside of a dome would be very unsafe for someone like Eddie, who likes to swing as high as possible. He would risk crashing into the curved walls of the dome.

Kari's bridge plan might have worked fine if she had used a stable foundation. Sand may be a nice cushion to fall down on, but it doesn't provide a solid base for the wooden posts that are holding the bridge up. The weight of people on the bridge would contribute to the downward pulling force on the end posts. If the posts aren't in a solid foundation like concrete that can withstand downward pressures, they might give in to the forces and begin to lean. Once that happens, the entire bridge would be at risk for collapse.

NOTE: The designs proposed in this account are fictitious. They are ideas presented for the exercise of applying building and design concepts to a real-life situation. Actual playground designs must meet strict federal standards and regulations.